PERCEPTIONS OF GMOS AT MEREDITH COLLEGE

By

Samantha C. Duerring

An Honors thesis submitted to the

Department of Biological Sciences

in partial fulfillment of the requirements for the degree of Bachelors of Science

Meredith College

Raleigh, North Carolina

April 30th, 2020

Samantha C. Duerring

Honors Student: Samantha C Duerring

Karthik Aghoram

Thesis Director: Dr. Karthik Aghoram

Jennifer D McMillen

Honors Director: Dr. Jennifer McMillen

Publication Agreement

I hereby grant to Meredith College the non-exclusive right to reproduce, and/or distribute this work in whole or in part worldwide, in any format or medium for non-commercial, academic purposes only.

Readers of this work have the right to use it for non-commercial, academic purposes as defined by the "fair use" doctrine of U.S. copyright law, so long as all attributions and copyright statements are retained.

Meredith College may keep more than one copy of this submission for purposes of security, backup and preservation.

Samantha Duerring
April 30th, 2020

Table of Contents

Acknowledgments

I would like to thank all the people who supported me throughout this process, my parents, all of my professors at Meredith, especially my research mentor Dr. Aghoram, and my friends Sarah H, Lilly W, and Alex R.

Abstract

From grocery stores to the evening news, the debate surrounding genetically modified organisms (GMOs) rages on. Some believe them to be part of humanity's future, while others view them as a hazard to consumer safety. The purpose of this study is to look at community concern surrounding GMOs, identify major causes for concern, and compare these correlations to trends found in literature. An online survey was sent to Meredith College faculty, students, staff, and alumni asking questions about overall concern about GMOs, education status, religious and personal beliefs, where survey takers got the majority of their scientific knowledge from, and the strength of their scientific background. Over the course of one month in 2019, the survey received 417 completed responses. In terms of overall concern, 52.09% of students (n=137) either were not at all concerned about GMOs overall or very slightly concerned, 22.05% were somewhat concerned, and 25.85% were concerned to extremely concerned. Faculty, staff, and alumni (n=70) mirrored this trend with 45.45% of them not at all concerned or very slightly concerned, 24.68% somewhat concerned, and 29.87% concerned to very concerned. These trends of lower overall concern about GMOs on Meredith's campus may be because the college promotes the pursuit of correct, valid, scientific knowledge and the importance of scientific research, and is an environment with educated individuals who have better access to scientifically verified information and know how to look for and access it.

Introduction

Since the phrase genetically modified organism (GMO) was coined, it has been a hot topic of debate and conversation in everything from scientific journals to social media. Some believe GMOs are the future of humanity, providing people with easily accessible, more wholesome foods; however, there are others who claim GMOs are indeed unsafe for consumption and can cause myriad of life-threatening diseases and environmental problems. One question that few people seem to be asking is why? For what reasons do GMOs have such a fiercely contested place in mainstream society? This paper will focus on the biological and sociological aspects of the debate, as well as explore the history behind genetically modified organisms, how they have influenced the development of today's society, why they may become necessary in the future, and the current acceptable uses of genetically modified organisms in society.

According to the United States Department of Agriculture, GMO's are "are organisms produced through genetic modification" (USDA, 2019). But what are they really? How does this seemingly clear-cut definition relate to the all-encompassing topic that has become public perceptions of GMO's? In order to fully understand the concept of genetically modified organisms, one must journey back in time to their very beginning (Appendix Figure 1). Since the 1970s, genetic modification has been thought of by many as something that occurs in a laboratory. Many believe that cutting-edge science and technology are used to create organisms that could never exist outside the bounds of human engineering; however, there is evidence to support the argument that humankind has been manipulating the genetic material of organisms for around 30,000 years (Rangel, 2015). The first example of this is thought to be the selective breeding of wolves into the modern-day dog. The traits that early humans desired were artificially selected for, eventually allowing the differentiation of the numerous dog breeds existing today. This same process was also utilized in regards to plants as far back in history as 7800 B.C.E in relation to the domestication of certain types of wheat in Asia (Rangel, 2015). This same type of genetic modification via artificial selection continues to this day, slowly but surely changing crops to the point of bearing little to no resemblance to their ancestral forms. One of the most prominent examples of this is corn. Its ancestor possessed small ears with very few kernels. Yet through hundreds of years of artificial selection developed the plant with large amounts of both (Rangel, 2015). Organisms that demonstrate genetic modification through artificial means are much more prolific than most people believe.

In the 1970s, one of the first major breakthroughs in what can be considered modern genetic engineering was discovered by Herbert Boyer and Stanley Cohen (Rangel, 2015). They were able to take genetic material from one type of bacteria and insert it into the genetic material of another, taking the first steps toward the modern type of genetic engineering usually thought of today as GMOs.

Some of the earliest GMO's came in the form of pest and herbicide tolerant varieties of soybeans, cotton, and corn (Wunderlich et al., 2015). One of the more well-known examples of this is *Bacillus thuringiensis*, more commonly known as BT corn, introduced into mainstream agriculture in 1995 (Hsaio, 2015). Another example would be Monsanto's glyphosate-tolerant, or "Roundup ready" crops (Wilkerson, 2015) Around the same time, the "Flavr Savr tomato" was introduced into stores, making it the GMO food to be widely available in the U.S. This tomato was designed to have a longer shelf life. According to Rangel, "these tomatoes were modified to include a DNA sequence that inhibited production of a natural tomato protein, increasing the firmness and extending the shelf life of the Flavr Savr variety" (Rangel, 2015). After a short period of success however, the production of these tomatoes was stopped due to the increasing worries of customers. One can surmise that this is most likely when the war on GMOs first began.

There are many reasons that average consumers seem to dislike genetically modified organisms, one of which is Biophilia. This term "refers to an innate predisposition to have an aesthetic appreciation of and attraction to natural environments" (Scott et al., 2018). The idea of naturalness is not necessarily safe or better, yet it is perceived as such by the average consumer. One must look no further than the nearest grocery store to locate examples of this. The labels of items are rife with sayings such as "all natural," and "no artificial flavors." Although naturalness seems very clear cut when put into these terms, its true definition is much murkier. Opinions on what is natural or unnatural vary greatly between person to person, yet the idea of contagion plays a major role in determining an object's level of naturalness (Rozin, 2005; Scott et al., 2018). Contagion is the belief that naturalness can be devalued upon coming into contact with something deemed foreign or unnatural. An example of this would be a fly landing in a person's food, or a plant coming into contact with genetic material from a separate source. Even less invasive processes, like adding extra Pulp to orange juice, could be considered unnatural to some. It is understandable then that through this lens genetically modified organisms would be perceived in a negative manner.

Another facet to the debate surrounding genetically modified organisms is morality. In most instances of this type of opposition to genetic engineering, religion is the main player. According to some people's beliefs, nature is sacred and should never be tampered with by humans (Paarlberg, 2014). According to this philosophy, genetically modified organisms should not be created even if the technology is at hand. In still other theologies, the scientists themselves are viewed as tampering with processes that should be left in the realm of the divine, playing god in a sense (Mallinson et al., 2018). It is thought that taking genetic material from one organism and placing it into another through artificial means is against the natural way the world works, especially if genetic material is transferred between separate species. An example would be a tomato containing a frost resistant gene originating in a species of fish. Since such fish to tomato gene transfer could typically not happen through natural means, such intervention could be seen as doing a task that should not be attempted by humans.

Every day the media plays a large role in forming people's perceptions on any number of topics, including genetically modified organisms. According to (Wunderlich & Gatto, 2015), the majority of consumers get their information about GMOs from sources like television and radio instead of scientific articles. Media can be a very useful resource for getting information out to large amounts of consumers; however, problems arise when accuracy of said information is questionable. Advertisement in supermarkets and restaurants also influences the perceptions of consumers. For example, the fast food chain Chipotle claims its food is "GMO-free" and the grocery store Whole Foods "promotes 'your right to know' and offers information about how to shop to avoid GMOs," illustrating the willingness of large businesses to use people's misgivings as advertising (Scott et al., 2018).

In both the United States and abroad, the various workings of government are major components of the debate surrounding genetically modified organisms. Policy surrounding regulations such as labeling and usage are not the same on an international scale, and differing opinions between consumers can possibly make creating such policies within individual countries more difficult. For example, there is evidence to support that in the United States, white males are more accepting of GMOs, and older vegetarian women are the least receptive to the concept (Mallinson et al., 2018). GMOs can also be used in some situations as a way of furthering separate agendas. This can be illustrated in Paarlberg (2014) by the behavior of activist groups. Oftentimes these groups are able to gain the trust of consumers and spread negative information that is not always based on scientific research. In such instances, consumers may not be getting enough information from all sides of the argument to allow them to make informed choices about government regulations of genetically modified organisms.

However, while this debate continues to rage, the planet is changing and with it only increased challenges that humans are predicted to experience in the not so distant future. This article seeks to not only explore the concept of genetic engineering, but also why it is necessary for the future of humanity as a species in regards to malnutrition and disease. In many developing countries, malnutrition is a problem. The world's population is increasing at a startling rate, and food in some areas becomes increasingly scarce. According to David Freedman, "The world will have to grow at least 70% more food in order to keep up with population growth" (2013). Genetically modified crops could be one of the ways that this increase in food production could come to pass. They could also be designed to increase crop yields and weather the effects of climate change via drought, flood, and other resistive properties (Freedman 2013). This will lower the price of food, while making it more plentiful. For example, Daniel Norero makes note of a group of Italian researchers who have published studies on genetically modified maize and its increase in crop yield over the past 20 years (Feb. 2018). One could assume that similar benefits could arise if other staple foods such as rice and legumes underwent similar modifications; then food scarcity may prove to be not as much of an issue.

Malnutrition of the developing world does not only involve quantity of food produced but quality as well. According to Norero, "Some 2 billion people suffer from some type of nutritional deficiency" (Jun. 2018). There have been efforts to try and circumvent this issue, such as supplementation via medical intervention and attempts to bolster food supplies; however, both of these avenues have been fraught with difficulty (Norero Jun. 2018). Selective breeding processes that are designed to create crops with higher and more specific nutrient content can be used, yet they are time consuming and must remain within a small group of sexually compatible species. This is not necessarily the case for genetically engineered crops (Norero Jun. 2018). Genetic material can be inserted into a plant from many more varied sources such as different plant species, viruses, and bacteria to name a few (Norero Jun. 2018). One of the most well known examples of this is a genetically modified species of rice infused with higher amounts of beta carotene, otherwise known as golden rice, which is a promising source for vitamin A. According to Norero, "Globally, the severe deficiency of this vitamin causes 500,000 cases of irreversible blindness, millions of cases of xerophthalmia, and up to 2 million deaths per year, most of these in children under 5 years of age" (Jun. 2018). This rice is using the genetic material from a strain of bacteria and maize to increase the levels of beta carotene found in the edible portion, which supplements beta carotene levels in a diet that would otherwise be very poor in regards to this vital nutrient. Unfortunately, due to public opposition and regulations against genetically modified organisms, this product has yet to be used commercially; however, with the human population predicted to rise substantially in the next century, such things may need to be re-evaluated in the future.

Another way in which genetically modified crops will come to be an even bigger necessity in the future is disease control and not necessarily in regards to humans. According to Erik Stokstad, bananas are an integral part of the lives of nearly 400,000,000 people worldwide, as well as a substantial export in some countries (2017); however, this most popular of fruits is under attack. In the 1950s a fungus decimated the Gros Michel variety, causing the industry to switch over to the more resistant Cavendish banana. Unfortunately, this strength is no longer the case. A related fungus to the one that decimated banana crops in the 1950s has managed to infect the Cavendish. This would not be as much of a problem if bananas were more genetically diverse; however, the plant reproduces asexually via rhizomes, leaving it vulnerable to disease (Ordonez et al. 2015). Currently there is no known fungicide that can kill the Fusarium wilt tropical race 4, more commonly known as TR4, and methods to prevent further contamination have only proved partially successful (Stokstad 2017). All is not lost, however, for a group of Australian researchers have discovered a gene, RGA2, from a wild banana variety. Upon inserting this gene into the Cavendish, its immunity to TR4 is showing to be expressed (Stokstad 2017). Despite this genetically modified banana not yet being used commercially, a similarly genetically modified crop has already been proven to work in similar circumstances. This involves the rainbow papaya, and how it saved the entirety of the Hawaiian papaya industry. From the 1940s to the 1990s, papaya ringspot virus (PRSV) plagued the Hawaiian Islands until the once thriving industry was on the edge of collapse (Davidson 2008). The virus, spread by aphids, compromised the plants photosynthetic abilities, eventually leading to death, if not for the work of a virologist named Dennis Gonsalves. He and his colleagues devised a strategy with the help of newly developed bio technologies to inoculate papayas with genetic material from a more mild form of the virus (Davidson 2008). This new resistant form of papaya was soon crossed with a species preferred by Hawaiian farmers, and thus today's rainbow papaya was born (Davidson 2008). Without this genetically modified crop, one could assume that the Hawaiian papaya industry would have most likely disintegrated, and as genetic diversity in many of today's most widely used plants continues to dwindle, it is possible that similar techniques in regards to genetic modification of crops will need to be applied now and in the future in order to sustain a growing population.

Based upon the evidence detailed above, one can surmise that genetically modified crops are important for the future of humanity. The world is changing, the population is growing, and the planet is getting warmer. Higher crop yields will be necessary to meet increasing demand, as will more nutritious options for the staples from which people live. Increased large scale agriculture and selective breeding will lead to less genetic diversity and plants more susceptible to disease. Despite the differing attitudes surrounding genetically modified crops, they may just prove to be one of the best ways to safeguard the future for generations to come.

Despite the apparent ferocity of the debate surrounding the topic of genetically modified organisms, they do in fact have some acceptable uses in society. According to a survey conducted by Widmar et al., the majority of people in the United States seem to be more accepting of GMO usage in the areas of medicine and public health (Widmar et al., 2017). One example in the aforementioned study involves the possible use of GM mosquitoes to help prevent spreading of the Zika virus (Widmar et al., 2017). Percentages of those in favor of GMO uses in medicine and health care were upwards of 50%, while other areas such as grain, produce, and livestock production were shown to be significantly lower (Widmar et al., 2017). Certain demographics also seem to play a part in acceptance; as mentioned before, survey participants that were white college-educated men appear to be more accepting of GMO's (Widmar et al., 2017). More recently however, a new twist on genetically modified food has been invented in the form of the impossible burger, released by the company Impossible Foods (Wolf, 2019). This burger contains no animal product yet is rumored to taste like actual meat. This is done by integrating hemoglobin originally produced via soybeans into the finished product; however, it was found that a form of yeast was able to produce the necessary proteins more effectively upon the use of recombinant DNA technology (Wolf, 2019). This form of yeast, *Pichia pastoris*, which has been used for formation of other genetically modified products is the only way impossible foods could make heme production efficient enough for large scale manufacturing (Wolf, 2019). One may surmise, based on the current debate, that such foods would most likely not be able to gain enough approval from the public to be sold, yet the Impossible Burger is sold at a number of fast food chains including, but not limited to Burger King (Wolf, 2019). One could consider such acceptance a step forward for those in favor of GMOs.

It is obvious that based on my research into the literature, that the debate on genetically modified organisms is multifaceted. There are mixed feelings in both this country and on the international front. I intend to administer a survey to the Meredith College campus in order to determine if such trends found in the major literature will be reflected on a smaller scale. I hypothesize that if I were to administer the survey, then the results would echo the trends found in the wider area of research. I also wish to determine if education and other such qualities play a major role in the acceptance of genetically modified organisms on the Meredith College campus. I am hopeful that some aspect of the survey can be used to better educate people in regards to genetically modified organisms and better prepare them for that maelstrom that is the current debate awaiting them.

Methods

This study was designed to identify perceptions of genetically modified organisms on the Meredith College campus. This was done via the use of an anonymous survey completed using the Qualtrics survey program. IRB exemption was obtained and the survey ran for a total of a month, with reminder emails sent to all individuals with a current association to Meredith College, including students, staff, and faculty. 500 responses were obtained overall. The survey contained a list of twenty-one questions in order to test in multiple areas involving such categories as knowledge biophilia, religion, and information awareness. See list as follows:

Q1: What is your age group?

Q2: What is your status at Meredith College? Pick the most relevant option (for example, if you are a staff member who is also a part-time student, pick "Staff.")

Q3: With what gender do you identify?

Q4: What is your political ideology?

Q5: In relation to GMOs in food, I would describe myself as [Extremely knowledgeable, Very knowledgeable, Moderately knowledgeable, Slightly knowledgeable, Not at all knowledgeable]

Q6: Where would you say you obtain **most** of your information about GMOs.

Q7: How concerned are you about GMOs being toxic to humans?

Q8: How concerned are you about the environmental effects of growing GMO foods (plants and animals)?

Q9: Based on your religious beliefs, how concerned are you about GMOs in food?

Q10: How concerned are you about being able to afford the cost of food made with GMOs?

Q11: How concerned are you about the nutritional value of food made with GMOs?

Q12: How concerned are you about the taste and flavor of food made with GMOs?

Q13: Overall, how concerned are you about food made with GMOs?

Q14: How concerned are you about GMOs being used to make medicines (e.g., insulin)?

Q15: How concerned are you about climate change?

Q16: Which of these foods do you consider to be **natural?** Check all that apply.

Q17: Indicate if the following statement is true or false: Non-GMO corn does not have genes, whereas GMO corn does.

Q18: When was the last time you took a course related to biology?

Q19: How do you think genetically modified organisms (GMOs) are viewed by the general public?

Q20: Do you think the opinion concerning genetically modified organisms (GMOs) on Meredith's campus is more positive, negative, or the same as the general public?

Q21: Do you think the opinion concerning genetically modified organisms (GMOs) on Meredith's campus is more positive, negative, or the same as the general public?

After completing the survey, different groups originally specified in the survey questions were grouped to give more accurate results since some respondent categories were very small. Levels of concern were also grouped to better describe and compare the survey results to national surveys.

Results

Upon completion of the survey, 500 responses were recorded, with 83 being incomplete or otherwise non-valid. In relation to overall concern about GMOs on the Meredith College campus as whole, 50.35% of participants were shown to have some level of concern surrounding GMOs (Appendix Table 1). The question of differing levels of concern between students and faculty was asked. I hypothesized that concern among faculty (n=154) would be lower than the concern among students (n=263). It happened that 54.55% of faculty were concerned compared to 47.9% of students (Appendix Table 2.1, 2.2). We wanted to see if age could be a possible variable when it came to overall concern about GMOs. When isolating for different age groups, 47.11% of respondents 30 and under (n=259) had some form of concern, compared to 55.7% in the 31-50+ age range (n=158) (Appendix Table 3.1-6).

To understand the trends in overall concern, we asked where respondents got the majority of their information on GMOs from. This was asked due to evidence supporting that media is often used to negatively impact the public's perception of GMOs (Scott et al., 2018). From our survey, respondents who got the majority of their information from Scientific Primary Resources, Science News, and College courses (n=164) had an overall lower concern than other groups, with only 34.76% of respondents having some form of concern (Appendix Table 4.1, 4.2, 4.8). This is compared to those who got their information from public media (including TV, radio, online news, and newspapers) (n=135), with 52.97% having some form of concern (Appendix Table 4.3, 4.4, 4.5, 4.9). Respondents that got the majority of their information from social media and/or their social circles were the most concerned, with 59.78% of respondents having some form of concern (Appendix Table 4.6, 4.7, 4.10, 4.11).

According to the survey done by the Pew Research Center, there are no significant differences in overall concern about GMOs based on political ideology (Pew Research Center, 2016). In this study, we found that 47.66% of Conservative respondents (n=82) had some form of concern (Appendix Table 5.1), while similarly, 45.46% of Liberal respondents (n=187) had some form of concern about GMOs (Appendix Table 5.2). Moderate, Independent, Libertarian (n=148) had the highest level of concern, with 55.74% of respondents having some form of concern (Appendix Table 5.3, 5.4). According to a study published in the *Annual Review of Nutrition*, religion and moral values are the biggest influencers of opinion about GMOs (Scott et al., 2018). Most respondents, 84.17%, responded that their religion did not have concerns with GMOs. We found that of the respondents whose religion had concern about GMOs (n=66), 95.46% of them had some overall concerns about GMOs (Appendix Table 6.1-5).

When asked about how knowledgeable they were about GMOs, 33.58% of respondents said they were very slightly to slightly knowledgeable, 44.12% said they were moderately knowledgeable, and 22.31% said they were very knowledgeable to extremely knowledgeable. Looking at how self-perception of knowledge influences opinions on GMOs, when asked how concerned they were of using GMOs in medicines like insulin, 45.71% of respondents with low knowledge of GMOs had some form of concern (Appendix Table 7.1, 7.2). The trend moves downward as 42.93% of respondents with moderate knowledge said they had some form of concern of GMO used in medicine (Appendix Table 7.3), while only 30.11% of respondents with a high level of knowledge said they had some level of concern

(Appendix Table 7.4, 7.5). This trend correlates to Mallinson's work, which demonstrates an increased level of science knowledge decreases fears and concerns surrounding GMOs (Mallinson, 2018)

To further look into understanding the correlation between level of knowledge and concerns about GMOs, this survey asked respondents to indicate if the following was true or false: "Non-GMO corn does not have genes, whereas GMO corn does." 94.24% answered false correctly, with only 24 respondents responding incorrectly with true. Of the respondents who answered incorrectly, 79.16% of them had some level of concern about GMOs, compared to the respondents who answered correctly, with 48.6% having some level of concern (Appendix Table 8.1, 8.2).

Of the respondents who answered incorrectly, the majority perceived themselves as having moderate knowledge about GMOs (33.33%) or slight to very little knowledge on GMOs (62.03%). Only 4.17% claimed to be very knowledgeable (Appendix Table 8.3). Of the respondents who answered correctly, 31.8% claimed to be slightly to very slightly knowledgeable, 44.78% claimed to be moderately knowledgeable, and 23.41% claimed to be very to extremely knowledgeable (Appendix Table 8.4). Of the respondents who answered correctly, 34.09% are either currently in a biology course, or have taken one within the past year. 26.46% have taken a biology course 2-5 years ago, and 39.44% took a biology course 6+ years ago (Appendix Table 8.6). Contrastly, of respondents who answered incorrectly, 37.49% are currently in a biology course or have taken one within the past year, 37.5% have taken one 2-5 years ago, and 25% have taken one 6+ years ago (Appendix Table 8.5). The respondents who answered incorrectly mainly got their information about GMOs from public media (50.01%, online news articles, television, newspapers, documentaries) and social media and/or family and friends (37.5%). Only 12.5% got their information from college courses, and none (0%) responded that they received their information from primary scientific literature or scientific journals (Appendix Table 8.7).

Discussion

The intention behind the survey was to assess whether trends regarding opinions of GMOs mimicked national trends, and whether qualities such as being on a college campus and access to scientific information had any influence in increased acceptance of GMOs on Meredith Campus. Overall, the trends are generally consistent, with about 50% of the respondents having some form of concern, and with concern decreasing with age and education level. Trends of level of concern amongst different political ideologies were the same, all showing about the same level of concern. It was interesting to see that for religion, the majority of respondents did not have any conflicting views between their religion and GMOs; however, amongst the minority of respondents who cited religious conflict, there was a sharp increase in the level of concern, mimicking national trends. The decrease in overall religious concern may be a reflection of the diverse student body and student's exposure to other worldviews on campus.

The most obvious trend seen that was mirrored was the level of scientific knowledge and source for information surrounding GMOs. Respondents who got the majority of their information from scientific sources had overall lower concern compared to those using public media and social media sources. This correlates with the literature and phenomenon seen with both marketing, media, and activists using misinformation to drive opinions. In terms of the science question asked in my survey, the majority of respondents answered correctly, which may be in part to more than a third of all respondents having taken a biology course within the past year, and more than half of respondents taking one within the past 5 years. Answering the science question incorrectly was correlated to both increased concern and getting the majority of source information from public or social media (12.5% college courses and 0% scientific sources).

This survey conducted on Meredith campus confirmed that respondents with higher education, science backgrounds, and those who access scientific resources overall have lower concerns about GMOs in food and medicine. Being able to access and understand the science involved in genetically modified organisms facilitates better acceptance of their use. Conversely, if someone is both unaware of the actual science, and is receiving their information from public and social sources that may be influencing their opinions for the sake of marketing, pseudoscience, or social movements, misconceptions and false information may easily sway their opinion. There is an overwhelming similarity between the national data about media usage and opinions on GMOs. With increased accessibility to scientific information, and an increased understanding of the science behind GMOs, being on a college campus and taking biology classes can influence more favorable opinions since understanding is in a way a form of acceptance.

There were some elements though that were not able to be determined by the survey on the Meredith College campus, one of which is gender demographics. It is shown in some of the literature that older vegetarian women are typically more skeptical of GMOs and younger white males are more accepting. Since Meredith has an undergraduate student body that is entirely female, this result would have to be considered non-valid. The issue of biophilia was also another aspect that could not be determined. The survey only had one question that yielded any data, although it was determined to be invalid because of the difficulty in defining the true definition of natural and what natural really means. The other difficulty in this survey was the sample size. Since there was a small pool of respondents compared to national surveys, many of the exposures were difficult to assess. This made it slightly more difficult to determine the truth behind some of the age values versus concern given that the respondents that were under 18 were very low and the majority of respondents were between the ages of 18 and 23. The smaller sample size also has a larger influence in the results. Since there were only a few respondents in certain categories (ex. only 4 under the age of 18, only 20-30 respondents for each source category), results were easily skewed because of the small sample size. The results had to be regrouped after the survey was complete to better observe trends amongst the different questions and demographics. However, this made it more difficult to see actual trends reflected in the study population. In order to truly observe trends in each of the original studied categories, this survey will have to be repeated with a larger sample size, and possibly run over multiple years.

In regard to further research on this project, I would like to see the survey run several more times at Meredith to see how many of the results can be duplicated. Some of the questions also could possibly be revised or divided into multiple questions to more accurately understand the respondent's opinions, such as the one dealing with biophilia. The survey could also possibly be extended to a larger sample audience to include other colleges and universities in the area, such as NC State and UNC Chapel Hill. The larger sample size will show better trends and will expand the demographics of students represented in the results. Eventually, the data from this survey along with further surveys could be used to further educate college students on the debate surrounding genetically modified, both encouraging and teaching students how to better access factual primary and secondary scientific information, as well as educating students about how genetically modified organisms are made, and their potential positive social and economic impact on the world. This will hopefully better allow students and the wider population to make their own decisions based on science rather than being misled by less reputable sources of information. It would also be interesting to mirror the national surveys with a national survey of college campuses to further look into the correlation that education, access to information, and diverse student populations have positive impacts on opinions surrounding GMOs.

Information and how it is portrayed is powerful. Despite the many benefits GMOs serve in both the local and international populations and for the security of health and sustenance for the future, the narrative of GMOs continues to be led with misinformation and the spread of fear. It is important for consumers to understand how to access and process information to better form their own opinions and maintain truthfulness in the public conversation. Through increasing accessibility to scientific information and providing education to people, making the biology more understandable, GMOs may become more accepted amongst a broader audience. However, when the loudest voices continue to drive opinions and social trends to their own benefits, both the science and the millions of people who will benefit from it continue to go unheard.

End References

Braimah, J. A., Atuoye, K. N., Vercillo, S., Warring, C., & Luginaah, I. (2017). Debated

agronomy: public discourse and the future of biotechnology policy in Ghana. *Global

bioethics = Problemi di bioetica*, 28(1), 3-18. doi:10.1080/11287462.2016.12616046.

Davidson S. 2008. Forbidden Fruit: Transgenic Papaya in Thailand. Plant Physiology.

147(2). https://www.ncbi.nlm.nih.gov/pmc/articles/PMC2409016/

Dorius, S. F., & Lawrence-Dill, C. J. (2018). Sowing the seeds of skepticism: Russian state

news and anti-GMO sentiment. *GM Crops & Food*, 9(2), 53–58. https://doi-

org.proxy108.nclive.org/10.1080/21645698.2018.1454192

Freedman D. 2013. The Truth about Genetically Modified Food. Scientific American.

https://www.scientificamerican.com/article/the-truth-about-genetically-modified-

food/

Ordonez N et al. 2015. Worse Comes to Worst: Bananas and Panama Disease—When Plant

and Pathogen Clones Meet. PLoS Pathogens 11(11).

https://journals.plos.org/plospathogens/article/file?id=10.1371/journal.ppat.1005197

&type=printable

Hsaio J. (2015). GMOs and Pesticides: Helpful or Harmful? Science in the News.

http://sitn.hms.harvard.edu/flash/2015/gmos-and-pesticides/

Mallinson, L., Russell, J., Cameron, D.D. et al. Food Sec. (2018) Why rational argument

fails the genetic modification (GM) debate. *Food Security*, 10: 1145.

https://doi.org/10.1007/s12571-018-0832-1

Norero D. Feb. 2018. GMO crops have been increasing yield for 20 years, with more

    progress ahead. Alliance for Science. Cornell.

    https://allianceforscience.cornell.edu/blog/2018/02/gmo-crops-increasing-yield-20-

    years-progress-ahead/

Norero D. Jun. 2018. Unfairly demonized GMO crops can help fight malnutrition. Alliance

    for Science. Cornell. https://allianceforscience.cornell.edu/blog/2018/06/unfairly-

    demonized-gmo-crops-can-help-fight-malnutrition/

Paarlberg, R. (2014). A dubious success: The NGO campaign against GMOs. *GM Crops &*

    *Foo*d, 5(3), 223–228. https://doi-

    org.proxy108.nclive.org/10.4161/21645698.2014.952204

Rangel, G. (2015). From corgis to corn: a brief look at the long history of GMO technology.

    In: Science in the News. http://sitn.hms.harvard.edu/flash/2015/from-corgis-to-corn-a-

    brief-look-at-the-long-history-of-gmo-technology/

Rozin, P. (2005). The Meaning of "Natural." *Psychological Science* (0956-7976), 16(8),

    652–658. https://doi-org.proxy108.nclive.org/10.1111/j.1467-9280.2005.01589.x

Scott, S. E., Inbar, Y., Wirz, C. D., Brossard, D., & Rozin, P. (2018). An overview of

    attitudes toward genetically engineered food. *Annual Review of Nutrition*, 38, 459–

    479. https://doi-org.proxy108.nclive.org/10.1146/annurev-nutr-071715-051223

Stokstad E. 2017. GM banana shows promise against deadly fungus strain. Science.

    American Association for the Advancement of Science.

    https://www.sciencemag.org/news/2017/11/gm-banana-shows-promise-against-

    deadly-fungus-strain

[USDA] United States Department of Agriculture. Agricultural Biotechnology Glossary.

https://www.usda.gov/topics/biotechnology/biotechnology-glossary

Wolf J. (2019). The Microbial Reasons Why the Impossible Burger Tastes So Good.

American Society for Microbiology. https://www.asm.org/Articles/2019/May/The-

Microbial-Reasons-Why-the-Impossible-Burger-Ta

Widmar NJO, Dominick SR, Tyner WE, Ruple A. (2017). When is genetic modification

socially acceptable? When used to advance human health through avenues other than

food. PLOS One.

https://journals.plos.org/plosone/article?id=10.1371/journal.pone.0178227

Wunderlich S, Gatto KA. (2015). Consumer Perception of Genetically Modified Organisms

and Sources of Information. Advances in Nutrition 6(6).

https://academic.oup.com/advances/article/6/6/842/4555145

Wilkerson J. (2015). Why Roundup Ready Crops Have Lost their Allure.

http://sitn.hms.harvard.edu/flash/2015/roundup-ready-crops/

Pew Research Center. (2016). Public opinion about genetically modified foods and trust in

scientists connected with these foods. Pew Research Center.

https://www.pewresearch.org/science/2016/12/01/public-opinion-about-genetically-

modified-foods-and-trust-in-scientists-connected-with-these-foods/

Appendix

Figure 1 (Page 4)

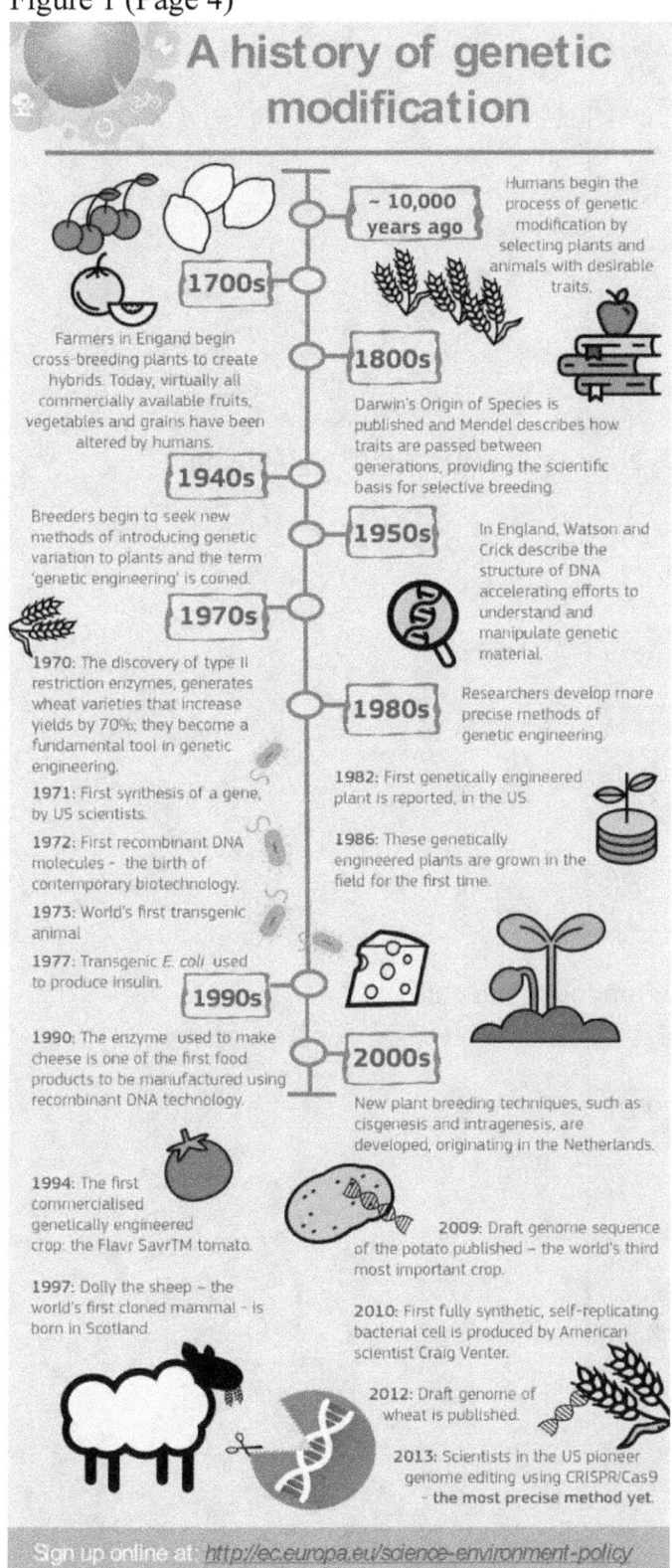

Table 1 Overall Level of Concern about GMOs amongst Respondents

Not at all concerned	21.58%
Very slightly concerned	28.06%
Somewhat concerned	23.02%
Concerned	19.18%
Extremely concerned	8.15%
Total responses	417

Table 2.1 Overall Concern about GMOs amongst Faculty, Staff, and Alumni

Not at all concerned	18.83%
Very slightly concerned	26.62%
Somewhat concerned	24.68%
Concerned	19.48%
Extremely concerned	10.39%
Total responses	154

Table 2.2 Overall Concern about GMOs amongst Students

Not at all concerned	23.1%
Very slightly concerned	28.90%
Somewhat concerned	22.05%
Concerned	19.01%
Extremely concerned	6.84%
Total responses	263

Table 3.1 Overall Concern about GMOs amongst Respondents Ages 18 and Under

Not at all concerned	25.00%
Very slightly concerned	25.00%
Somewhat concerned	0.00%
Concerned	50.00%
Extremely concerned	0.00%
Total responses	4

Table 3.2 Overall Concern about GMOs amongst Respondents Ages 18 to 21yrs old

Not at all concerned	25.14%
Very slightly concerned	25.14%
Somewhat concerned	22.40%
Concerned	19.67%
Extremely concerned	7.65%
Total responses	183

Table 3.3 Overall Concern about GMOs amongst Respondents Ages 22 to 30 yrs old

Not at all concerned	25.00%
Very slightly concerned	34.72%
Somewhat concerned	25.00%
Concerned	12.50%
Extremely concerned	2.78%
Total responses	72

Table 3.4 Overall Concern about GMOs amongst Respondents Ages 31 to 40 yrs old

Not at all concerned	25.64%
Very slightly concerned	33.33%
Somewhat concerned	17.95%
Concerned	12.82%
Extremely concerned	10.26%
Total responses	39

Table 3.5 Overall Concern about GMOs amongst Respondents Ages 41 to 50 yrs old

Not at all concerned	12.20%
Very slightly concerned	21.95%
Somewhat concerned	31.71%
Concerned	24.39%
Extremely concerned	9.76%
Total responses	41

Table 3.6 Overall Concern about GMOs amongst Respondents Ages 50 and older

Not at all concerned	12.82%
Very slightly concerned	29.49%
Somewhat concerned	21.79%
Concerned	23.08%
Extremely concerned	12.82%
Total responses	78

Table 4.1 Overall Concern amongst Respondents whose primary source of information on GMOs was Specialized Primary Scientific Literature

Not at all concerned	45.95%
Very slightly concerned	24.32%
Somewhat concerned	5.41%
Concerned	10.81%
Extremely concerned	13.51%
Total responses	37

Table 4.2 Overall Concern amongst Respondents whose primary source of information on GMOs was Popular Scientific Articles

Not at all concerned	34.38%
Very slightly concerned	25.00%
Somewhat concerned	9.38%
Concerned	18.75%
Extremely concerned	12.50%
Total responses	64

Table 4.3 Overall Concern amongst Respondents whose primary source of information on GMOs was Online News Articles

Not at all concerned	11.36%
Very slightly concerned	23.86%
Somewhat concerned	36.36%
Concerned	23.86%
Extremely concerned	4.55%
Total responses	88

Table 4.4 Overall Concern amongst Respondents whose primary source of information on GMOs was Television

Not at all concerned	13.04%
Very slightly concerned	17.39%
Somewhat concerned	39.13%
Concerned	21.74%
Extremely concerned	8.70%
Total responses	23

Table 4.5 Overall Concern amongst Respondents whose primary source of information on GMOs was Newspapers and Magazines

Not at all concerned	19.05%
Very slightly concerned	28.57%
Somewhat concerned	28.57%
Concerned	14.29%
Extremely concerned	9.52%
Total responses	21

Table 4.6 Overall Concern amongst Respondents whose primary source of information on GMOs was Food Documentaries

Not at all concerned	16.22%
Very slightly concerned	18.92%
Somewhat concerned	27.03%
Concerned	27.03%
Extremely concerned	10.81%
Total responses	37

Table 4.7 Overall Concern amongst Respondents whose primary source of information on GMOs was Social Media

Not at all concerned	9.09%
Very slightly concerned	30.30%
Somewhat concerned	30.30%
Concerned	18.18%
Extremely concerned	12.12%
Total responses	33

Table 4.8 Overall Concern amongst Respondents whose primary source of information on GMOs was College Courses

Not at all concerned	22.22%
Very slightly concerned	46.03%
Somewhat concerned	19.05%
Concerned	9.52%
Extremely concerned	3.17%
Total responses	63

Table 4.9 Overall Concern amongst Respondents whose primary source of information on GMOs was Radio

Not at all concerned	00.00%
Very slightly concerned	66.67%
Somewhat concerned	33.33%
Concerned	0.00%
Extremely concerned	0.00%
Total responses	3

Table 4.10 Overall Concern amongst Respondents whose primary source of information on GMOs was Conversations with Family Members

Not at all concerned	9.09%
Very slightly concerned	40.91%
Somewhat concerned	13.64%
Concerned	31.82%
Extremely concerned	4.55%
Total responses	22

Table 4.11 Overall Concern amongst Respondents whose primary source of information on GMOs was Other

Not at all concerned	34.62%
Very slightly concerned	15.38%
Somewhat concerned	19.23%
Concerned	23.08%
Extremely concerned	7.69%
Total responses	26

Table 5.1 Overall Concern amongst Respondents whose political ideology is Conservative

Not at all concerned	17.07%
Very slightly concerned	29.27%
Somewhat concerned	26.83%
Concerned	20.73%
Extremely concerned	6.10%
Total responses	82

Table 5.2 Overall Concern amongst Respondents whose political ideology is Liberal

Not at all concerned	26.74%
Very slightly concerned	27.81%
Somewhat concerned	21.93%
Concerned	17.11%
Extremely concerned	6.42%
Total responses	187

Table 5.3 Overall Concern amongst Respondents whose political ideology is Moderate/Independent

Not at all concerned	17.14%
Very slightly concerned	29.29%
Somewhat concerned	22.86%
Concerned	20.71%
Extremely concerned	10.00%
Total responses	140

Table 5.4 Overall Concern amongst Respondents whose political ideology is Libertarian

Not at all concerned	25.00%
Very slightly concerned	0.00%
Somewhat concerned	12.50%
Concerned	25.00%
Extremely concerned	37.50%
Total responses	8

Table 6.1 Overall Concern amongst Respondents whose religious views have Extreme Concern related to GMOs

Not at all concerned	0.00%
Very slightly concerned	0.00%
Somewhat concerned	0.00%
Concerned	25.00%
Extremely concerned	75.00%
Total responses	4

Table 6.2 Overall Concern amongst Respondents whose religious views have Concern related to GMOs

Not at all concerned	0.00%
Very slightly concerned	0.00%
Somewhat concerned	15.79%
Concerned	63.16%
Extremely concerned	21.05%
Total responses	19

Table 6.3 Overall Concern amongst Respondents whose religious views have Somewhat Concern related to GMOs

Not at all concerned	0.00%
Very slightly concerned	6.98%
Somewhat concerned	41.86%
Concerned	37.21%
Extremely concerned	13.95%
Total responses	43

Table 6.4 Overall Concern amongst Respondents whose religious views have Very Slight Concern related to GMOs

Not at all concerned	3.70%
Very slightly concerned	29.63%
Somewhat concerned	37.04%
Concerned	25.93%
Extremely concerned	3.70%
Total responses	27

Table 6.5 Overall Concern amongst Respondents whose religious views have No Concern related to GMOs

Not at all concerned	27.47%
Very slightly concerned	32.72%
Somewhat concerned	20.06%
Concerned	13.58%
Extremely concerned	6.17%
Total responses	324

Table 7.1 Level of Concern regarding GMOs used in Medicine (i.e. insulin) amongst Respondents with Very Slight Self Perceived Knowledge of GMOs

Not at all concerned	16.67%
Very slightly concerned	29.17%
Somewhat concerned	25.00%
Concerned	20.83%
Extremely concerned	8.33%
Total responses	24

Table 7.2 Level of Concern regarding GMOs used in Medicine (i.e. insulin) amongst Respondents with Slight Self Perceived Knowledge of GMOs

Not at all concerned	25.00%
Very slightly concerned	31.03%
Somewhat concerned	28.45%
Concerned	10.34%
Extremely concerned	5.17%
Total responses	116

Table 7.3 Level of Concern regarding GMOs used in Medicine (i.e. insulin) amongst Respondents with Moderate Self Perceived Knowledge of GMOs

Not at all concerned	32.07%
Very slightly concerned	25.00%
Somewhat concerned	19.02%
Concerned	14.67%
Extremely concerned	9.24%
Total responses	184

Table 7.4 Level of Concern regarding GMOs used in Medicine (i.e. insulin) amongst Respondents with Self Perceived Knowledge of Very Knowledgeable of GMOs

Not at all concerned	50.00%
Very slightly concerned	17.57%
Somewhat concerned	14.86%
Concerned	8.11%
Extremely concerned	9.46%
Total responses	74

Table 7.5 Level of Concern regarding GMOs used in Medicine (i.e. insulin) amongst Respondents with Self Perceived Knowledge of Extremely Knowledgeable of GMOs

Not at all concerned	73.68%
Very slightly concerned	5.26%
Somewhat concerned	10.53%
Concerned	0.00%
Extremely concerned	10.53%
Total responses	19

Table 8.1 Level of Concern regarding GMOs among Respondents who answered True/False question on genetics Incorrectly

Not at all concerned	4.17%
Very slightly concerned	16.67%
Somewhat concerned	58.33%
Concerned	12.50%
Extremely concerned	8.33%
Total responses	24

Table 8.2 Level of Concern regarding GMOs among Respondents who answered True/False question on genetics Correctly

Not at all concerned	22.65%
Very slightly concerned	28.75%
Somewhat concerned	20.87%
Concerned	19/59%
Extremely concerned	8.14%
Total responses	393

Table 8.3 Level of Self Perceived Knowledge regarding GMOs among Respondents who answered True/False question on genetics Incorrectly

Very slightly knowledgeable	20.83%
Slightly knowledgeable	41.67%
Moderately knowledgeable	33.33%
Very knowledgeable	4.17%
Extremely knowledgeable	0.00%
Total responses	24

Table 8.4 Level of Self Perceived Knowledge regarding GMOs among Respondents who answered True/False question on genetics Correctly

Very slightly knowledgeable	4.83%
Slightly knowledgeable	26.97%
Moderately knowledgeable	44.78%
Very knowledgeable	18.58%
Extremely knowledgeable	4.83%
Total responses	393

Table 8.5 Last taken Biology Course by Respondents who answered True/False question on genetics Incorrectly

Current semester	20.83%
Past semester	8.33%
Past year	8.33%
Past 2 to 5 years	37.50%
Past 6 to 10 years	0.00%
Over 10 years ago	25.00%

Table 8.6 Last taken Biology Course by Respondents who answered True/False question on genetics Correctly

Current semester	17.81%
Past semester	6.36%
Past year	9.92%
Past 2 to 5 years	26.46%
Past 6 to 10 years	6.62%
Over 10 years ago	32.82%

Table 8.7 Primary Source of Information about GMOs for Respondents who answered True/False question on genetics Incorrectly

Specialized primary scientific literature	0.00%
Popular science articles	0.00%
Online news articles	29.17%
Television	12.50%
Newspapers and magazines	4.17%
Food documentaries	4.17%
Social media	16.67%
College courses	12.50%
Radio	0.00%
Conversations with family members	12.50%
Other	8.33%
Total responses	24

www.ingramcontent.com/pod-product-compliance
Lightning Source LLC
Chambersburg PA
CBHW080423190526
45161CB00004B/265